9.92

PLANTS
AND
US

Angela Royston

Heinemann Library
Des Plaines, Illinois

Designed by AMR Ltd.
Printed and bound in Hong Kong/China by South China Printing Co. Ltd.

03 02 01 00 99
10 9 8 7 6 5 4 3 2 1

Library of Congress Cataloging-in-Publication Data

Royston, Angela.
      Plants and us / Angela Royston.
         p. cm. – (Plants)
      Includes bibliographical references and index.
      Summary: Surveys the many uses of plants, including food, drink,
spices and herbs, creams and perfumes, medicine, wood and paper,
clothes, and decoration for gardens and parks.
      ISBN 1-57572-825-7 (lib. bdg.)
      1. Human-plant relationships—Juvenile literature. 2. Plants,
Useful—Juvenile literature. [1. Plants, Useful.] I. Title.
II. Series: Plants (Des Plaines, Ill.)
QK46.5.H85R69 1999
581.6'3—dc21   98-42810
                                              CIP
                                              AC

Acknowledgments
The Publishers would like to thank the following for permission to reproduce photographs:
Ardea: A. Paterson p. 21; Liz Eddison: pp. 13, 16; Garden and Wildlife Matters: pp. 7, 12, 26, 27,
J. Hoare p. 20, S. North p. 8; Chris Honeywell: pp. 11, 14,18, 22, 28, 29; Oxford Scientific Films:
J. McCammon p. 24, Oxapia p. 9; Science Photo Library: J. Howard p. 23; The Stock Market: p. 6;
Tony Stone Images: W. Curtis p. 15, N. Dolding p. 17, M. Gowan p. 10, G. Haling p. 19, R. Torrez
p. 5, P. Tweedle p. 4; Trip: J. Hurst p. 25.
Cover photograph: Paul Chesley, Tony Stone Worldwide
Every effort has been made to contact copyright holders of any material reproduced in this book.
Any omissions will be rectified in subsequent printings if notice is given to the Publisher.

Any words appearing in bold, **like this**, are explained in the Glossary.

Contents

We All Need Plants   4

Air Fit to Breathe   6

Vegetables   8

A Field of Grain   10

Fruit and Nuts   12

Drinks   14

Spices and Herbs   16

Creams and Perfumes   18

Medicine   20

Wood and Paper   22

Clothes   24

Gardens and Parks   26

Changing Color   28

Plant Map   30

Glossary   *31*

Index   *32*

More Books to Read   *32*

We All Need Plants

All animals and people rely on plants for food. Even animals that don't eat plants eat animals that do. This giraffe is eating the leaves of a thorny bush.

People grow fields of corn and other food plants. They also use plants for other things. This house is built of wood from trees.

Air Fit to Breathe

The air contains **oxygen**. All living things breathe in oxygen and breathe out **carbon dioxide**.

During the day, plants take in carbon
dioxide through their leaves and turn
it into oxygen. Plants put fresh oxygen
we need to breathe back into the air.

Vegetables

We eat plants because they contain the
vitamins and **minerals** our bodies need
to grow and stay healthy. Vegetables
can be **roots**, **stems**, leaves, or **flowers**.

Carrots and potatoes are swollen
roots. We eat the leaves of cabbage
and lettuce and the flowers of
cauliflower and broccoli.

A Field of Grain

Farmers all around the world grow fields of rice or wheat. This farmer is **harvesting** rice by hand. We eat the **seeds** called grains.

When wheat is cut, the grains are collected and **ground** into a powder called flour. Flour is used to make bread, pasta, and cakes.

Fruit and Nuts

The fruit is the part of the plant that contains **seeds** that could grow into new plants. Apples, oranges, peaches, and strawberries are all fruits.

Each fruit has one or more seeds
surrounded by sweet, juicy flesh. Nuts
are seeds too, but they are surrounded
by a hard shell.

Drinks

Aside from milk and water, most of
what we drink comes from plants.
Some fruits are squeezed to make juice.

The leaves of tea plants are picked by hand before they are dried and made into tea. Coffee beans are roasted and **ground** before they are used.

Spices and Herbs

We add herbs and spices to food to
make it taste better. The leaves of these
herbs have a strong taste and smell.

Spices are made from the **roots** and
bark of plants that grow mainly in
hot countries. Many spices are **ground**
into powder before they are sold.

Creams and Perfumes

All of these **cosmetics** have been made from plants. The labels tell you which plants have been used.

Flowers with a strong, sweet smell are made into perfumes. These lavender flowers may be used to make soap or powder smell nice.

Medicine

In the past, most medicines came from plants. Today plants are still used to treat some illnesses. This little rosy periwinkle helps to treat leukemia, a disease that affects people's blood.

The **bark** of the cinchona tree contains a drug called quinine. This drug is used to treat an illness called malaria, which is common in some hot countries.

Wood and Paper

All of these things are made from wood. The table is wood also. Some kinds of trees are specially grown so that we can use their wood.

The wood of some trees is mashed down and made into huge rolls of paper for newspapers, books, packaging, and other things.

Clothes

Plants can be made into clothes. These
fluffy **seeds** of cotton are spun into
thread and woven into cotton cloth.

Many plants have very strong colors that are used to make dyes. These clothes are being dyed many different colors.

Gardens and Parks

Plants improve our lives in other
ways too. Gardens and parks give us
somewhere peaceful to relax and
enjoy ourselves.

These **rainforest** plants are growing in a **national park** in Central America. We must protect all plants, not just the ones we use today.

Changing Color

You can use beets to make a dye, but you must ask an adult to help you. Find two or three beets and have an adult chop them. Put them in a pan of water. Add a small piece of white cotton cloth.

Ask the adult to boil the water on top
of the stove for about a half an hour.
Make sure that the water does not
boil away. Let the water cool. What
color is the cloth now?

Plant Map

A Strawberry Plant

leaf

flower

fruit

stem

roots

An Oak Tree

bark

leaves

roots

trunk

Glossary

bark	tough outer layer that protects the trunk of a tree
carbon dioxide	a gas that is made when living things breathe out
cosmetics	creams, lotions, and powders that people use on their skin and hair
flower	the part of a plant that makes new **seeds**
ground	broken up into a powder
harvesting	gathering crops, like wheat, when they are fully grown
minerals	substances found in the earth that plants and animals need to stay healthy
national park	area of land that has been put aside for plants or animals to live without being disturbed by people
oxygen	a gas that all living things need to breathe to survive
rainforest	rainy place where many trees and plants grow together
roots	parts of a plant that take in water, especially from the soil
seed	a seed contains a tiny plant and a store of food before it begins to grow
stem	the part of a plant from which the leaves and **flowers** grow
vitamins	special kind of foods that animals and people need to stay healthy

Index

carbon dioxide 6–7

coffee 15

cosmetics 18–19

cotton 24

dyes 25, 28–29

farmers 10

fruit juice 14

medicine 20–21

minerals 8

oxygen 6–7

paper 23

rainforest 27

tea 15

vitamins 8

wheat 10–11

wood 5, 22–23

More Books to Read

Butler, Daphne. *Gathering Food.* Chatham, NJ: Raintree Steck-Vaughn Publishers. 1995.

Curtis, Neil & Peter Greenland, *How Paper Is Made.* Minneapolis, MN: Lerner Publishing Group. 1992.

Fowler, Allan. *It's a Fruit, It's a Vegetable, It's a Pumpkin.* Minneapolis, MN: Children's Press. 1995.

Hughes, Meredith S. & E. Thomas Hughes. *Buried Treasure: Roots & Tubers.* Minneapolis, MN: Lerner Publishing Group. 1998.

Lewin, Betsy. *Walk a Green Path.* New York: Lothrop, Lee & Shepard Books. 1995.